W9-DHI-179

Austin, Texas

THE WORLD'S CLIMATES

COLD CLIMATES
DRY CLIMATES
EQUATORIAL CLIMATES
TEMPERATE CLIMATES

Published by Raintree Steck-Vaughn Publishers, an imprint of Steck-Vaughn Company

Library of Congress Cataloging-in-Publication Data
Lye, Keith.
Temperate climates / Keith Lye.
p. cm.—(The world's climates)
Includes bibliographical references and index.
Summary: Describes the weather, plant and animal life, and how people live in temperate regions around the world.
ISBN 0-8172-4827-7
1. Temperate climates—Juvenile literature.
[1. Temperate climates. 2. Climatology.]
I. Title. II. Series: Lye, Keith. World's climate.
QC993.75.L94 1997
551.6912—dc20 96-31162

Printed in Italy. Bound in the United States.
1 2 3 4 5 6 7 8 9 0 01 00 99 98 97

Cover picture: Sourwood tree along Unicoi Lake, Unicoi State Park, Georgia. Tony Stone Images.

Picture acknowledgments
Carol Kane 19, 32–33, 36; FLPA 16, 28; Robert Harding Picture Library 1, 5, 9, 11, 13, 15, 16–17, 19, 20, 21, 22, 23, 25, 26, 27, 31, 32–33, 34–35, 37, 38, 39, 40, 41, 42, 43; Trip 5, 7, 11, 17, 27, 31, 35, 38–39.

All illustrations Timothy Lole except p39, Alistair Wilson

Contents

The Middle Latitudes

CLIMATIC REGIONS

Climate is the usual, or average, weather of a place. It determines what plants and animals are found in an area, and it influences how people live. The two main factors that determine climate are temperature and precipitation. Precipitation includes rain, snow, sleet, hail, dew, and frost—in fact, all forms of moisture that come from the air.

Places with temperate climates occur in the middle latitudes—between the low latitudes of the hot tropics and the high latitudes of the cold poles. These places have four distinct seasons, so they do not suffer the extreme cold of polar and subarctic climates, nor do they have the constant, year-round heat of tropical regions. The precipitation is generally plentiful, so temperate regions do not include dry climates, even though

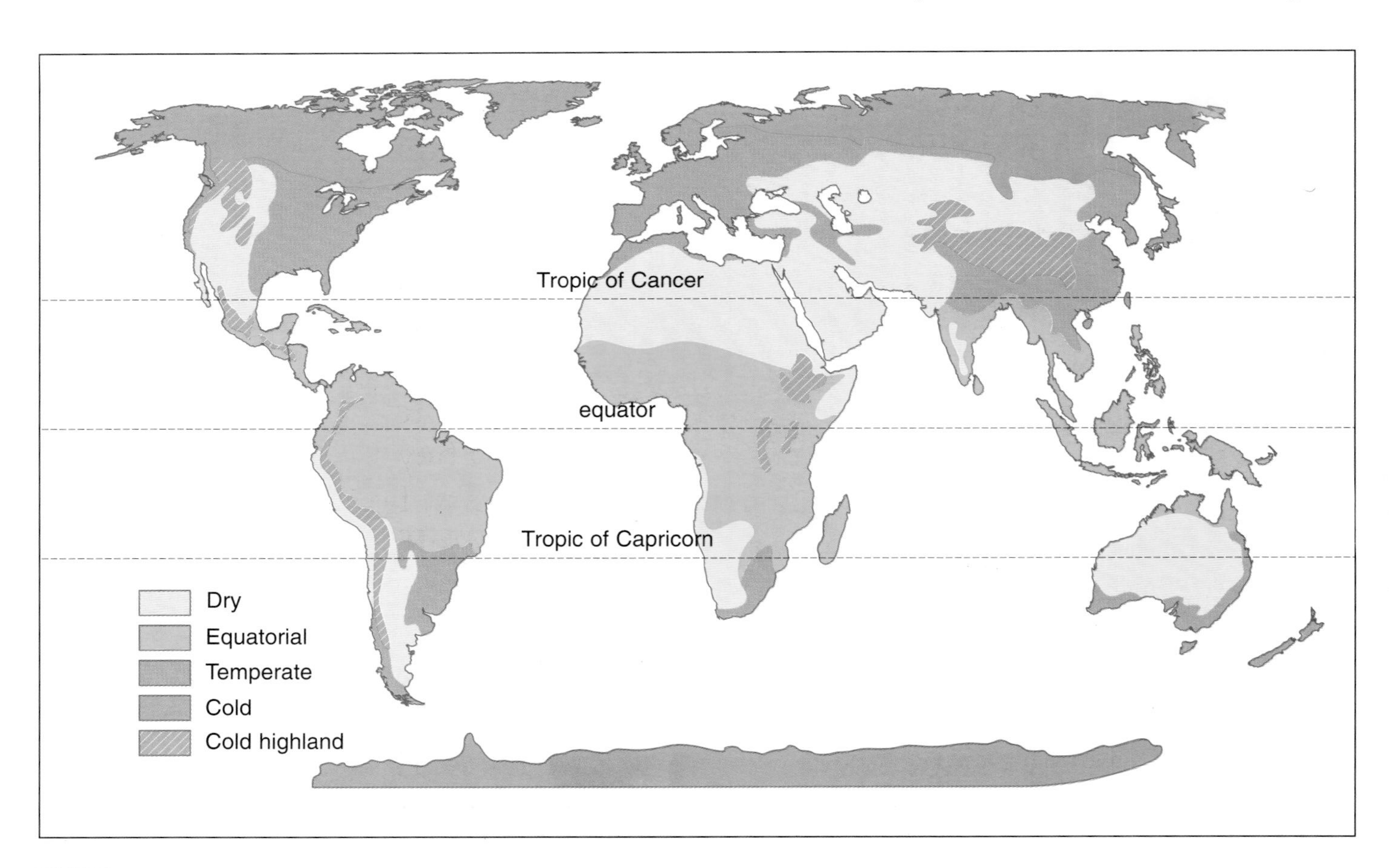

Above *Sunbathers on Bondi Beach, Sydney, Australia*

Right *Red Square, Moscow, in the winter*

Left *The main climate regions of the world*

some deserts and dry grasslands do occur in the middle latitudes.

There are three main kinds of temperate climates. The first includes places with hot, dry summers and mild, wet winters. These places are said to have a Mediterranean climate, after the kind of climate that occurs around the Mediterranean Sea. Mediterranean climates also occur in smaller areas in North and South America, Australia, and South Africa.

The second type of temperate climate is the rainy temperate climate. It has rain throughout the year. This climate occurs in the southwestern United States, much of eastern China, and southern Brazil.

The third type of temperate climate is the cold temperate climate. Places with this climate have cold winters and hot summers. They occur in the hearts of continents, far from the moderating effect of the sea. For example, the town of Stirling in Scotland lies on roughly the same latitude as Moscow in Russia. Yet the average temperature in Stirling in January is 37°F (3°C), while in Moscow it is 16°F (–8°C).

FOUR SEASONS

The seasons in temperate regions are caused by the tilt in the earth's axis. The axis is the imaginary line through the earth that joins the North Pole, the center of the earth, and the South Pole. As the earth travels around the sun, areas of the earth tilt toward the sun, getting more sunlight, while other parts tilt away, getting less sunlight.

On March 20 or 21, the sun is directly overhead at the equator (the imaginary line that runs around the earth, halfway between the two poles and divides the world into the Northern and Southern hemispheres).

March 20 or 21 is the spring equinox in the Northern Hemisphere. At the spring equinox, night and day are roughly 12 hours long all over the world.

After March 21, the Northern Hemisphere tilts increasingly toward the sun, until, on June 20 or 21, the sun is overhead at the Tropic of Cancer (latitude $23\frac{1}{2}°$ north). This is the summer solstice in the Northern Hemisphere.

After June 21, the Northern Hemisphere starts to tilt away from the sun.

By September 22 or 23, the sun is again overhead at the equator. This is the autumnal equinox in the Northern Hemisphere. At the autumn equinox, night and day are again roughly 12 hours long all over the world.

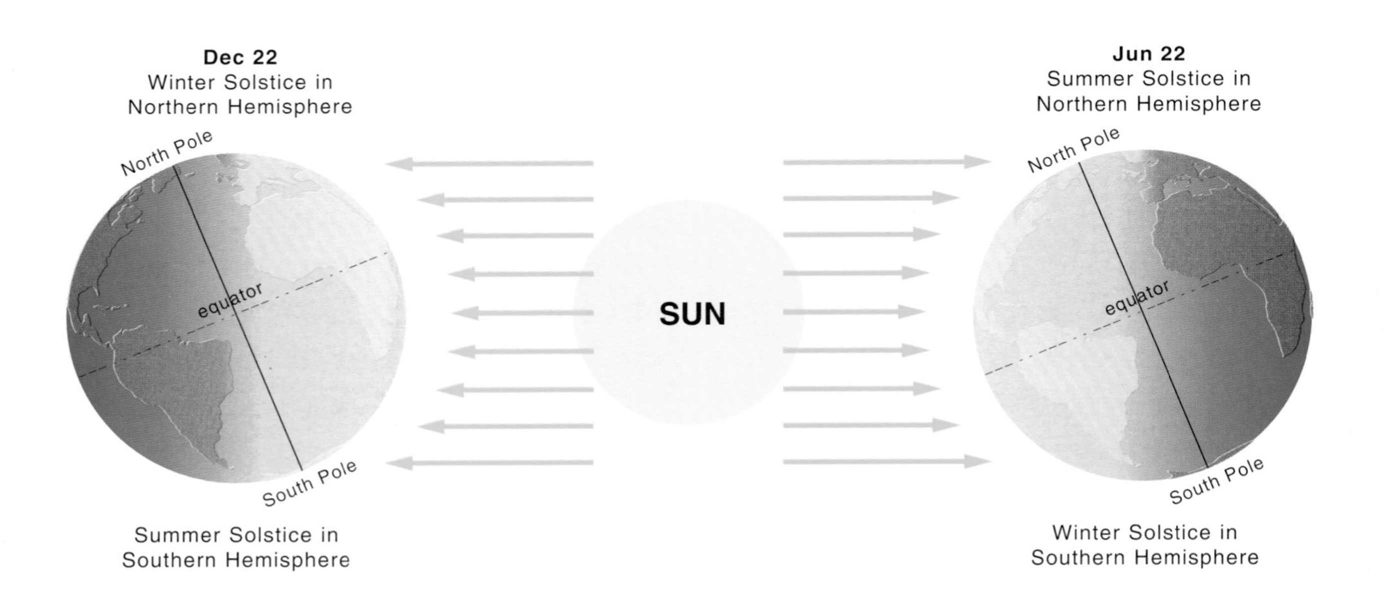

How the sun affects the summer and winter seasons

After September 23, the Northern Hemisphere starts to tilt away from the sun.

Finally, on December 20 or 21, the sun is overhead at the Tropic of Capricorn (latitude $23\frac{1}{2}°$ south). This is the winter solstice in the Northern Hemisphere.

But December 20 or 21 is the summer solstice in the Southern Hemisphere. The seasons are reversed south of the equator.

The same scene in summer (left) and winter (right)—old packhorse bridge at Carrbridge, Scotland

DEPRESSIONS AND ANTICYCLONES

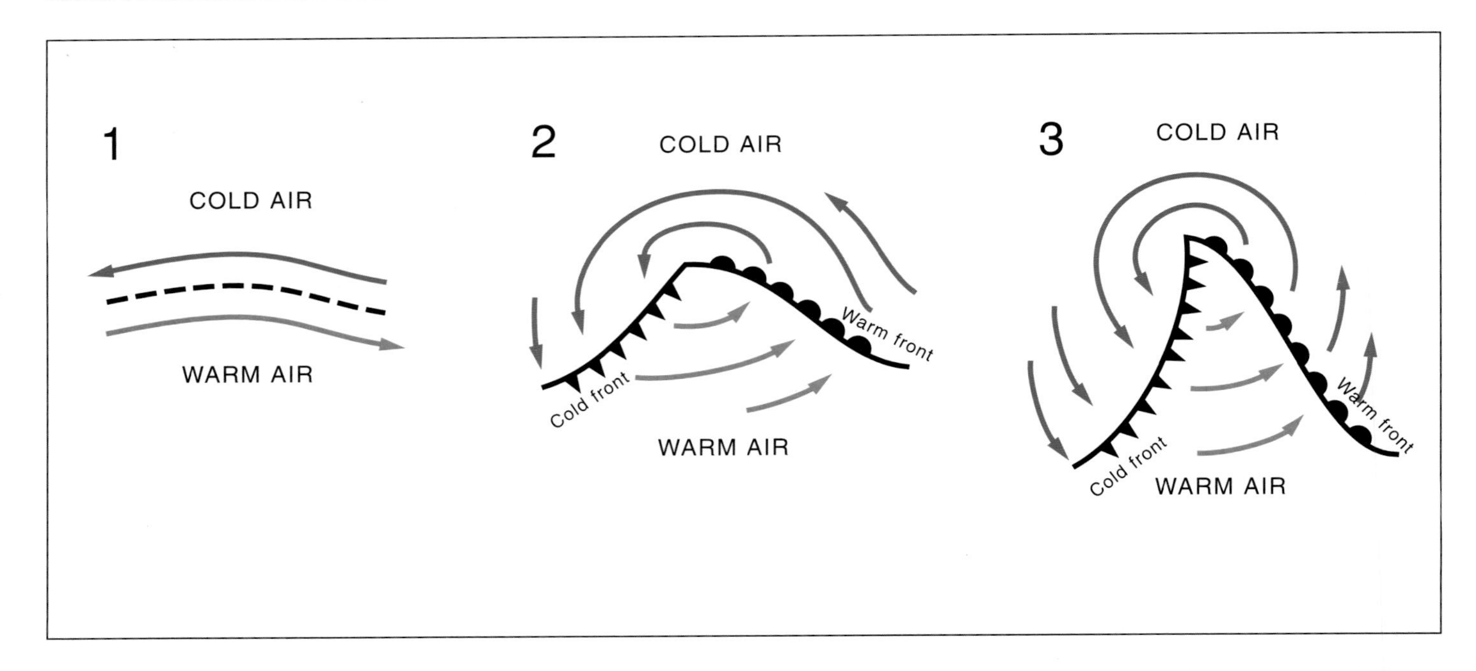

The weather changes a great deal in places that have a temperate climate, such as Great Britain. The reason for the changes is that the climate is influenced by two weather systems: anticyclones and depressions.

Anticyclones are regions of high air pressure where air is sinking. They bring stable weather. In the winter, anticyclones bring long spells of cold, frosty weather. In the summer, they bring blue skies and sunshine. By contrast, depressions are regions of low air pressure where warm air rises, clouds form in the rising air, and rain or snow are common.

Warm westerly winds, blowing from subtropical zones, meet up

Above *Depressions form along a polar front.*

Right *A sunny day and a cloudless sky—the result of an anticyclone—Brighton, England*

Below *This map of an anticyclone shows how the highest air pressures are at the center, with the winds blowing outward.*

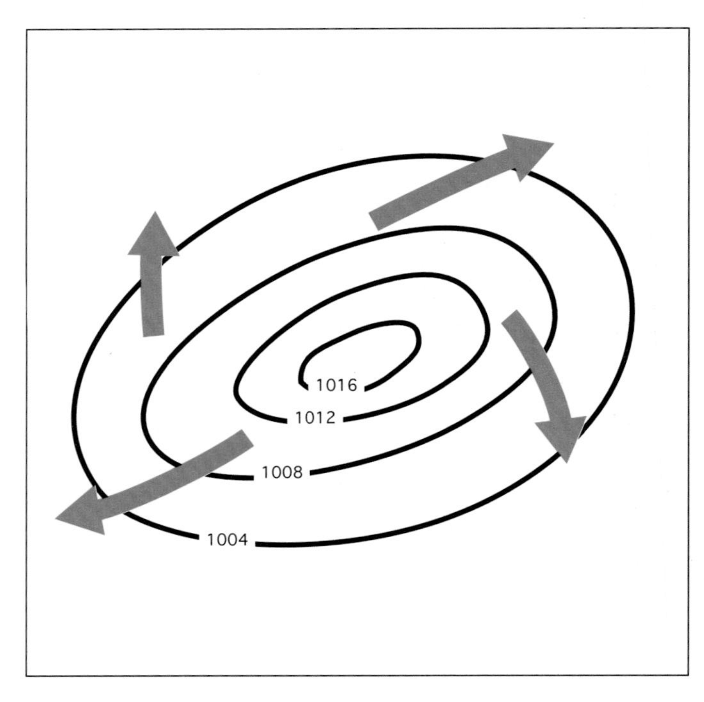

with cold easterly winds blowing from polar regions at the "polar front." This is how depressions form. Because warm and cold air do not mix easily, the warm air flows into bends in the polar front. Cold air flows in behind the warm air. This sets up a circular air system, or depression. At the center of the depression is the lowest air pressure.

So depressions are marked by fronts. The warm front is the edge of the advancing warm air, and the cold front is the edge of the cold air flowing in behind.

Along both fronts are bands of clouds, which bring stormy weather. The rain that falls during depressions is called cyclonic rain.

Eventually, the cold, heavy air overtakes the warm, light air. The places where this happens are called occlusions. Along occlusions, clouds and rain persist for some time.

When, eventually, the warm air becomes cold, the depression dies out.

CHANGEABLE WEATHER

As depressions move across the land, they bring changing weather conditions. Depressions vary greatly, measuring from 100 miles to 1,800 miles (150–3,000 km) or more across.

They often move rapidly from west to east across the temperate regions in the Northern Hemisphere. Many move at about 600 miles (1,000 km) a day, but some are slow moving or almost stationary.

Warm air contains invisible water vapor. When warm air rises, it cools. Cold air cannot hold as much water vapor as warm air can. So the water vapor condenses, turning into tiny ice crystals or water droplets (or both). Billions of ice crystals or water droplets (or both) form clouds.

The first sign of an approaching depression may be wispy, high clouds (cirrus), which form as the warm air rises above the cold air. Cirrus clouds are made up of tiny ice crystals.

Other clouds follow. Just ahead of the warm front are dark gray, flattish clouds (nimbostratus). These bring steady rain or snow.

When the warm front arrives, the rain eases. Temperatures rise and the air becomes more humid.

However, after a few hours (or a few days if the depression is moving slowly) the cold front arrives.

Along the cold front, which moves faster than the warm front, the cold air undercuts the warm. This pushes the warm air upward, and high clouds form in the rising air, creating thunderclouds

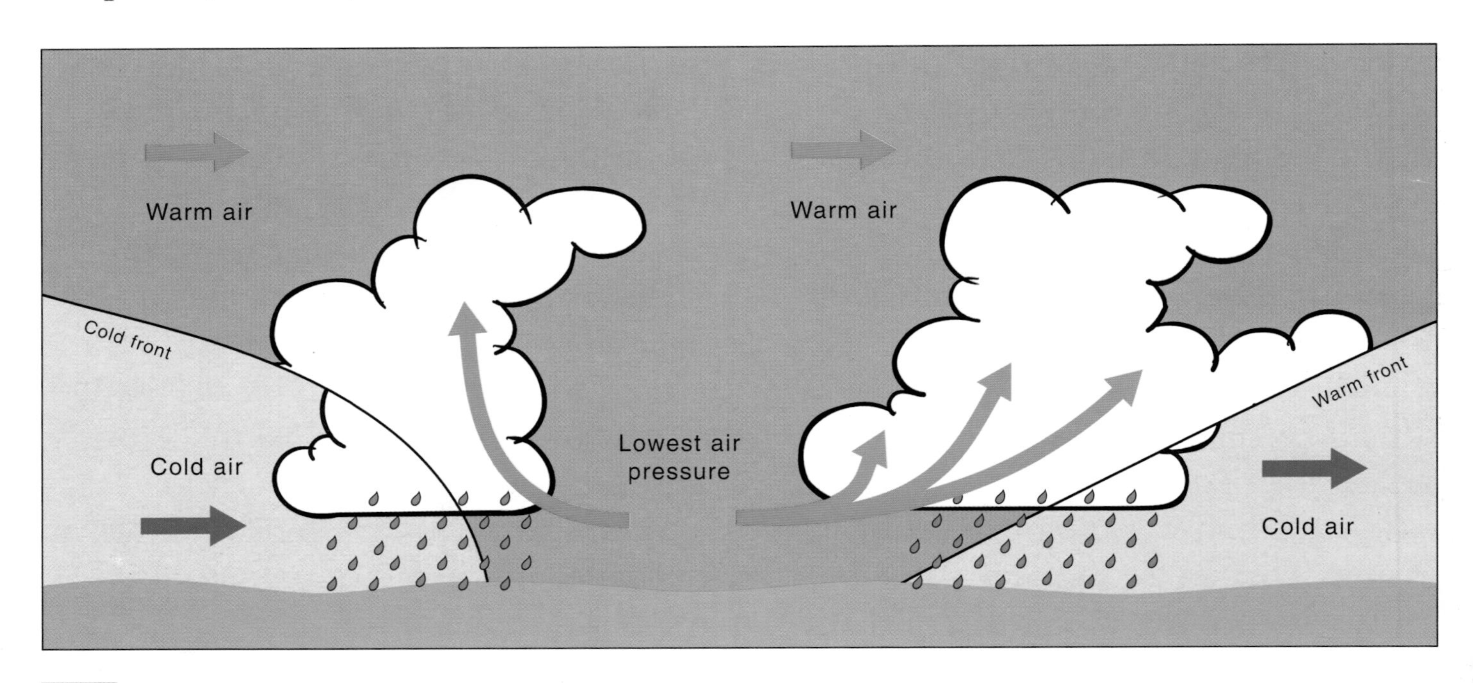

(cumulonimbus). Thunder and lightning, heavy rain and hail, or sleet and snow in the winter are features of thunderclouds.

These storms do not normally last for long and the cold front soon passes away. The skies clear and temperatures fall.

Above *This weather station supplies data that is used to produce weather forecasts.*

Left *A diagram of a depression, which is moving to the right as you see it*

Right *Heavy rain has caused the flooding of this athletic field.*

OCEAN CURRENTS AND CLIMATE

Ocean currents are the movements of water in the oceans. Onshore winds that blow across the currents are either warmed or chilled, depending on whether the current is warm or cool. This affects the climate of land near the sea.

Warm currents that flow from the tropics toward the polar regions warm them. Cold currents that flow from polar regions to the tropics cool them.

The Gulf Stream, a current that starts in the Gulf of Mexico, has a great influence on European climates. When this warm-water current emerges through the Strait of Florida, it flows up the east coast of the United States and then across the Atlantic.

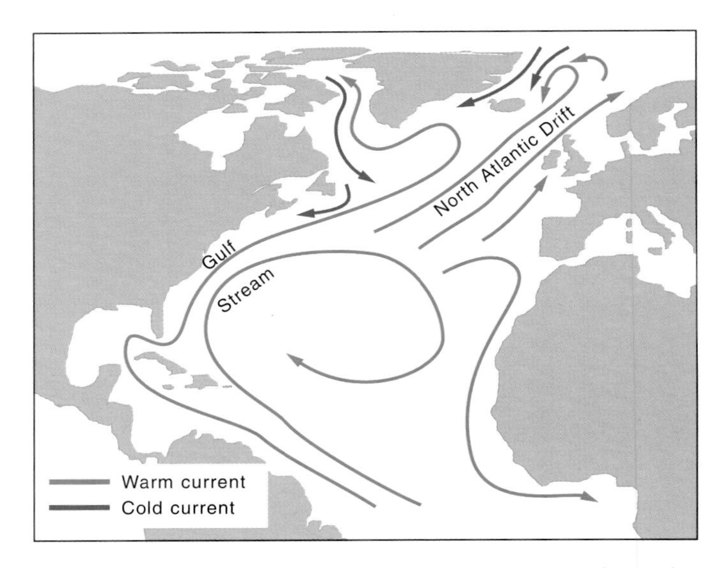

Above *Diagram showing the warm and cool currents of the North Atlantic Ocean*

Part of the current turns back down the coast of Africa as the Canary Current.

Another part forms a slower current called the North Atlantic Drift. This affects the climate of

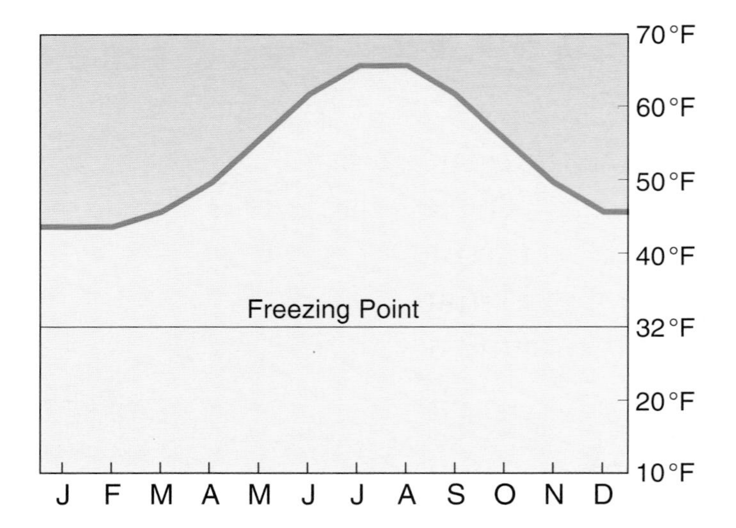

The warming effect of the Gulf Stream means that the winters in Brest, France, are mild, as this temperature graph shows.

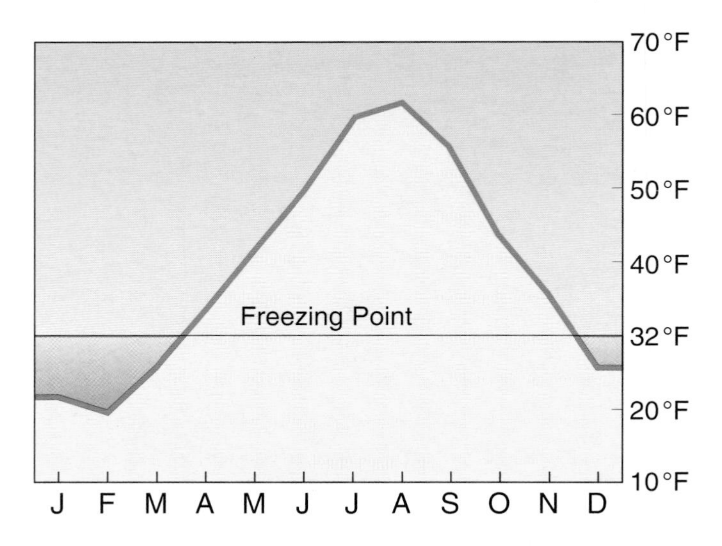

St. John's, Newfoundland, farther south than Brest, is not warmed by ocean currents as Brest is, and so has harsh winters.

Above *The port of Narvik, Norway remains ice free throughout the year because of the warm ocean current that flows nearby.*

such places as Iceland, the British Isles, and Scandinavia. It keeps most of Norway's seaports ice free throughout the year. Even the port of Hammerfest, 285 miles (460 km) north of the Arctic Circle, is always ice free.

The influence of the Gulf Stream and the North Atlantic Drift also explains why the climate of northwestern Europe is much milder than the climate of eastern Canada. For example, Brest in France is farther north than St. John's in Canada, but Brest has much warmer winters than St. John's. The climates of the east and west coasts of continents are often very different; warm currents usually flow along the east coasts of continents, while cold currents normally flow along the west coasts.

Mediterranean Climates

DRY SUMMERS, WET WINTERS

For summer vacations, many people want to find a beach resort where they can be sure that the sun will shine for most of the time. Many popular tourist resorts are in places with Mediterranean climates. That is because summers are sunny and dry, with average temperatures rising to 70 °F (21 °C) or, in some cases, over 80 °F (28 °C). The mild winters are another attraction, especially to retired people who grew up in countries with snowy winters. In Mediterranean climates, the average temperature in the coldest month is from 40 °F to 50 °F (6–10 °C).

Mediterranean climates occur in most of the lands around the Mediterranean Sea, although the coasts of Egypt and Libya have desert climates. Mediterranean climates also occur in smaller areas in central California, central Chile, the southwestern tip of South Africa, and parts of southern Australia. These regions have much in common. They are all known for their orange and lemon groves, vineyards, wine-making industries, and tourist beaches. In most areas, the original forests have been largely cut down and replaced by low scrub or open heath. Fire is a constant danger, especially during the dry summers.

Regions with Mediterranean climates have an average annual rainfall of from 16 inches to 36 inches (400–900 mm). Most of the

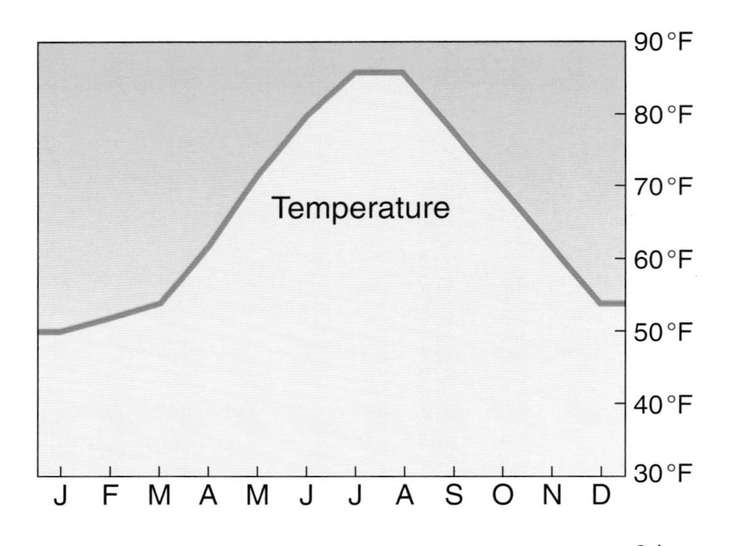

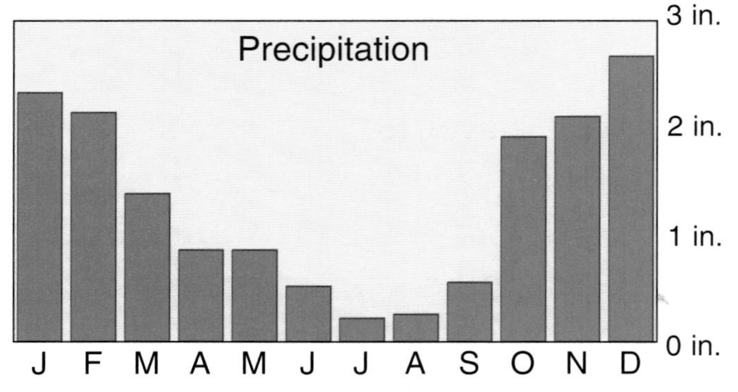

These climate graphs for Athens, Greece show its dry, warm summers and mild winters.

Above *Benidorm in Spain is very crowded during the summer.*

Left *The Mediterranean climate of Western Australia is suitable for vineyards such as this one.*

rain comes in the winter from depressions that move into the area.

In the summer, Mediterranean climate regions come under the influence of high air-pressure systems. When the land is intensely heated and hot air rises in strong currents, water vapor in the rising air condenses (turns into droplets of water) and forms thunderclouds that bring heavy rain.

However, storms do not last long, and afterward it is usually pleasantly cool.

PLANTS AND ANIMALS

Forests of mainly evergreen trees once grew over most of the land in Mediterranean climate regions. Mild winters enable plants to grow throughout the year without a winter pause in growth. But, from ancient times, the original forests were cut down. Many people around the Mediterranean Sea were sailors who needed lumber to build boats. Other people cleared forests to create farmland. In many areas, goats and other animals overgrazed the land, killing off the natural vegetation. Over large areas, fire has also restricted plant growth.

Some woodlands of chestnut and evergreen oak, including the valuable cork oak, still survive

Left *Olive trees are evergreens —this one is in Israel.*

Below *Wild horses live in the Camargue nature reserve in southern France.*

Below far right *Flamingoes are also part of the wide variety of life in the Camargue.*

around the Mediterranean Sea, but scrub or heath is now the main vegetation. The original forests in other regions with Mediterranean climates have also largely disappeared and have been replaced by scrub. The scrub has many names. In France it is called the maquis or garigue, in Italy the macchie, in California the chaparral, in Australia the mallee, and in South Africa the fynbo. The scrub is dominated by shrubs, such as the proteas of South Africa and Australia, the sagebrush of California, and the myrtles and rockroses of Europe.

Animal life is rich. For example, roe deer are widespread in parts of Mediterranean Europe. However, many other animals, such as lynxes and wolves, are now rare. Jackrabbits and pumas live in California's chaparral. Kangaroos are the commonest grazers in Australia. Duikers and hyraxes live in the South African fynbo.

The Mediterranean Sea region has many snakes and other reptiles. There are a wide variety of lizards. Birds include many species of warblers. The region is also the winter home of many birds that breed in the colder parts of northern Europe and then migrate. Smaller mammals include such creatures as dormice, foxes, hedgehogs, and rabbits.

THE MEDITERRANEAN SEA REGION

Mediterranean climates affect about 1.7 percent of the world's land area. The largest single area is in the Mediterranean Sea basin, a region that has been occupied by people since ancient times. It gave birth to some great civilizations, including the Minoan, ancient Greek, and Roman. Much of the land has been greatly affected by human activities.

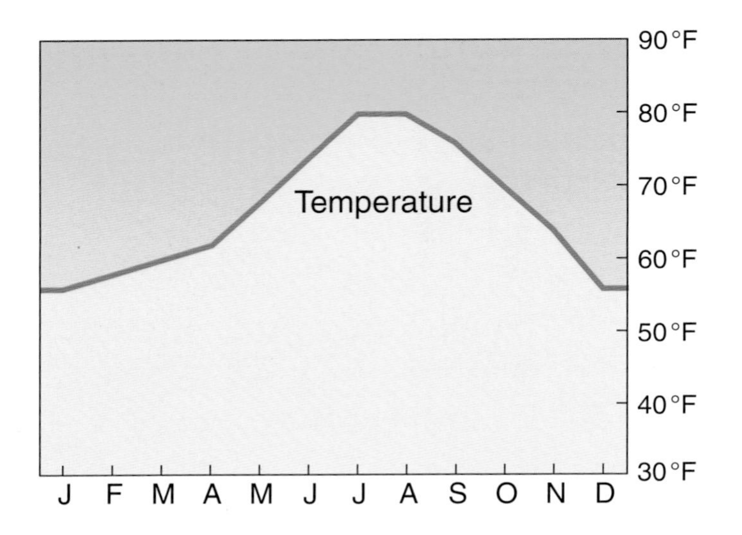

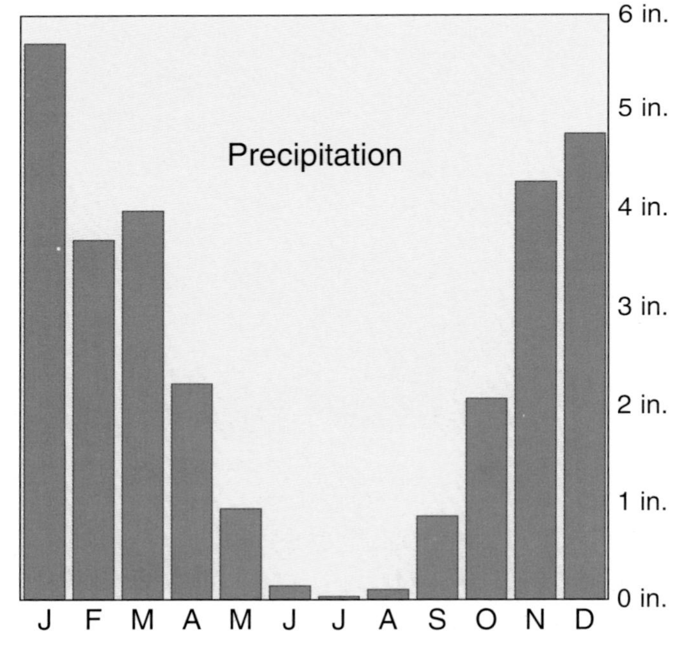

Climate graphs show Gibraltar's mild winter.

Apart from the interior uplands and the dry coasts of northeastern Africa, the lands around the Mediterranean Sea typically have hot, dry summers and mild, wet winters.

The effect of the winter depressions, which come from the Atlantic Ocean, is greater in the western countries. Generally speaking, rainfall decreases from west to east throughout the Mediterranean Sea.

The wetter areas have some woodland, although most trees are low or stunted. The typical vegetation consists of a dense cover of low bushes, including laurel and myrtle. Drier areas have sparser vegetation. Here grasses, with their shallow roots, struggle to survive during the hot summers. Common plants include the sweet-smelling broom, lavender, rockrose, rosemary, and thyme.

In the winter, cold winds blow in from the interior of Europe. For example, the bora brings cold weather to northern Italy and the east coast of the Adriatic Sea.

Above *Vineyards in Tuscany, Italy, flourish with hot, dry summers and mild, wet winters.*

Right *Even in the winter, the climate in Athens is mild.*

In France, the mistral is a strong wind that blows south down the Rhône valley, bringing cold and dry weather to the Rhône delta. Hot winds blow from the Sahara. The sirocco is a dry, dusty wind at the Algerian coast, but it picks up moisture over the sea, and when it reaches Sicily and southern Italy, it is hot and unpleasantly humid.

THE AMERICAS

The central valley and southern coastal regions of California and central Chile in South America both have Mediterranean climates.

However, the coastal regions in both California and Chile have much cooler summers than inland areas. They are cooled by onshore winds that blow over cold ocean currents. These cool winds also bring sea fogs.

The inland valleys, which run north to south behind coastal ranges, are much warmer. For example, Sacramento, in the central valley of California, has an average temperature of 73 °F (23 °C) in July. This is compared with 60 °F (15 °C) in San Francisco, which is on the coast.

The cool onshore winds sometimes affect the central valley when they are drawn inland through gaps in the coastal mountains.

In these regions, winters are mild. Sacramento has an average January temperature of 46 °F (8 °C).

However, nights are often cool. Frosts occasionally occur, and fires are sometimes lit in California orange groves to prevent frost damage. The rainfall is moderate, most of it coming from depressions that cross the regions in the winter.

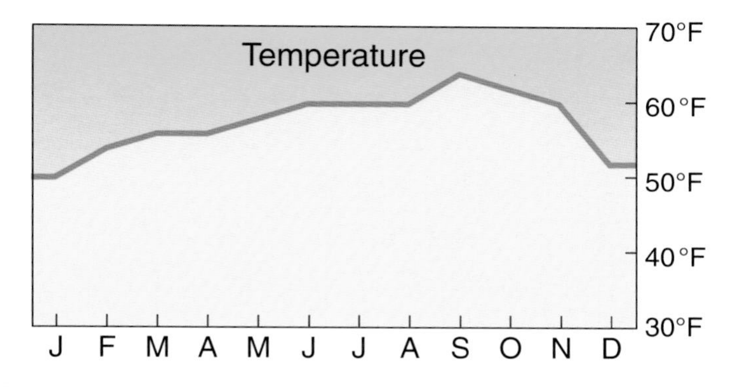

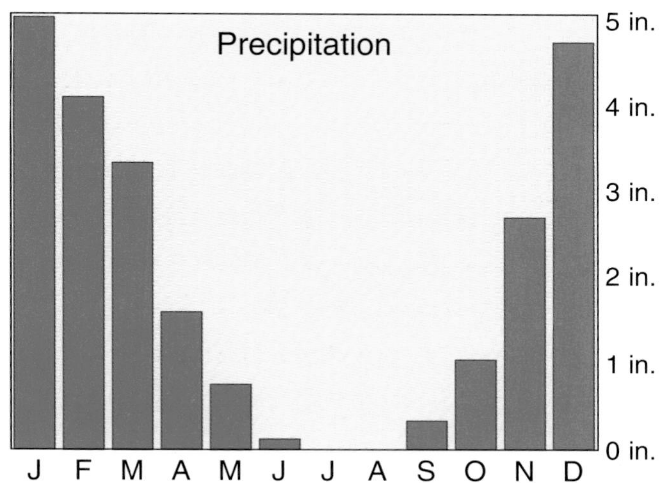

Both central California and central Chile are important farming areas that produce such crops as citrus fruits and grapes.

There are also areas of scrub, and wild lilac, highland live oak, and sumacs (plants of the cashew family) are common plants in the California chaparral. Redwood trees and yellow pines grow on the wetter northern coastal ranges of California, while forests of Araucaria pine grow on the slopes of the Andes. In parts of both California and central Chile, the Mediterranean regions merge into areas of steppe or desert with dry climates.

Above *Climate graphs for San Francisco show mild coastal summers that are caused by sea fog.*

Left *The California coast is one of the foggiest parts of the United States. Here The Golden Gate Bridge, San Francisco, is shrouded in fog.*

Right *A farmer tends his crops in Chile, South America.*

SOUTH AFRICA AND AUSTRALIA

Mediterranean climates occur in three other areas. All of them are well known for producing Mediterranean crops such as citrus fruits, wheat, grapes, and quality wines.

These regions include the southern tip of South Africa, around the city of Cape Town. Most rain comes in the winter from depressions, though occasional thunderstorms bring rain in the summer. The area has warm summers, with average temperatures reaching 70°F (21°C) in Cape Town in January. Winters are mild and the average temperature in July (the coldest month) is 54°F (12°C). Occasionally a warm wind, called the berg wind, reaches the area from the interior. It can raise temperatures to 100°F (38°C).

Two other regions with Mediterranean climates are in southern Australia. They include the southwestern tip of Australia, around Perth, while a second area includes the Adelaide district and part of Victoria. Between these two regions is a dry area bordering the Great Australian Bight. The winter depressions that bring rain to Australia's Mediterranean climate areas pass south of the Bight.

Southwestern Australia has a mild, almost frost-free climate. The average

Right *Protea, the national emblem of South Africa, grows in the Mediterranean climate of the Cape of Good Hope.*

Right *A sunny day in Perth, Western Australia. During the winter, depressions bring rain to Perth, making it even wetter than Adelaide.*

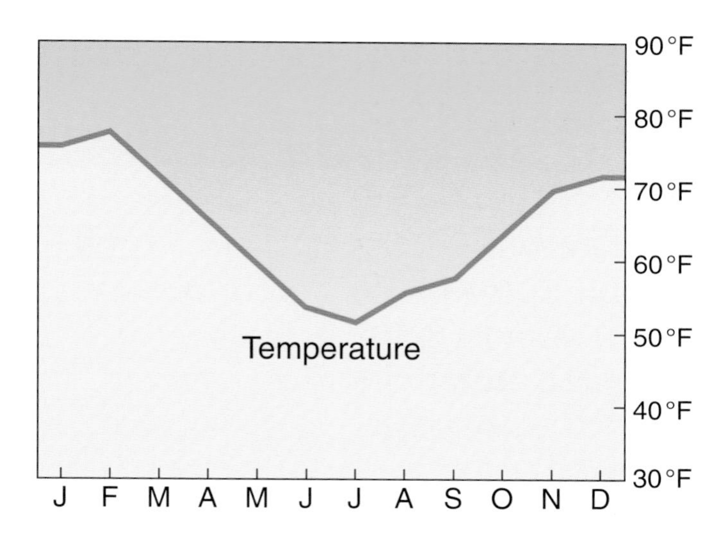

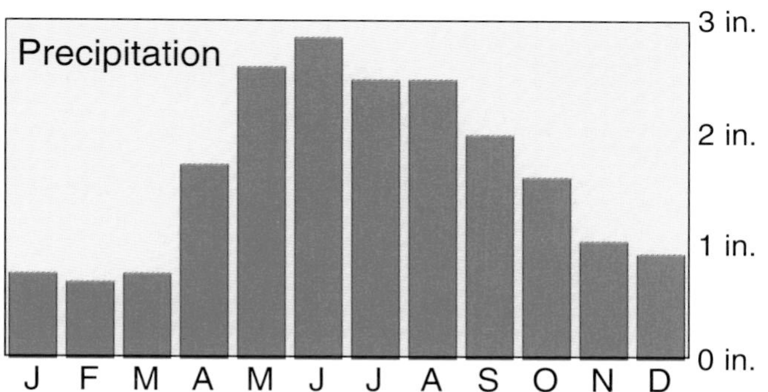

Climate graphs for Adelaide show rainfall throughout the year, with most in winter.

temperature in Perth in the coldest month (June) is 55 °F (13 °C). Perth has an average annual rainfall of about 35 inches (880 mm), most of which is brought by depressions in the winter, between May and August. Adelaide is drier. Some rain falls in summer, but the winter months are still the wettest.

The plant life in these regions changes from the wetter coastal areas to the drier inland areas.

In southwestern South Africa, the remains of coniferous forests still grow, but fynbo scrub is the main kind of natural vegetation.

In Australia, the wetter coastal areas contain some forest, but mallee scrub (a dwarf eucalyptus tree) is dominant inland.

Rainy Temperate Climates

WET AND WARM

There is another group of temperate climatic regions that has rain throughout the year. This group is divided into three main types, classified according to their temperatures in summer. One group has hot summers, another has warm summers, and a third has cool summers.

Temperate rainy climates with hot summers occur mostly on the eastern sides of continents, where the coastal areas are influenced by warm ocean currents. These regions are often described as subtropical, even though they lie outside the tropics, because they have mild or warm winters. Rain occurs in every month, but summer is often the wettest season. Much of the rainfall occurs during thunderstorms, when intense heating of the land causes warm air to rise and then drop its moisture. Hurricanes also bring rain in the summer, but in the winter rain is often caused by depressions. The southeast coast of Queensland, Australia, around the city of Brisbane, has a climate of this type. Other areas include the southeastern United States, much of China, southern Japan, part of South America (including southern Brazil), and South Africa's southeast coast.

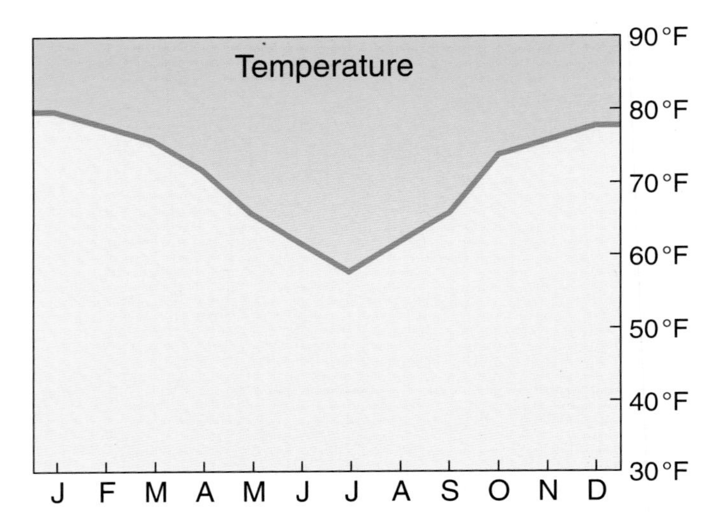

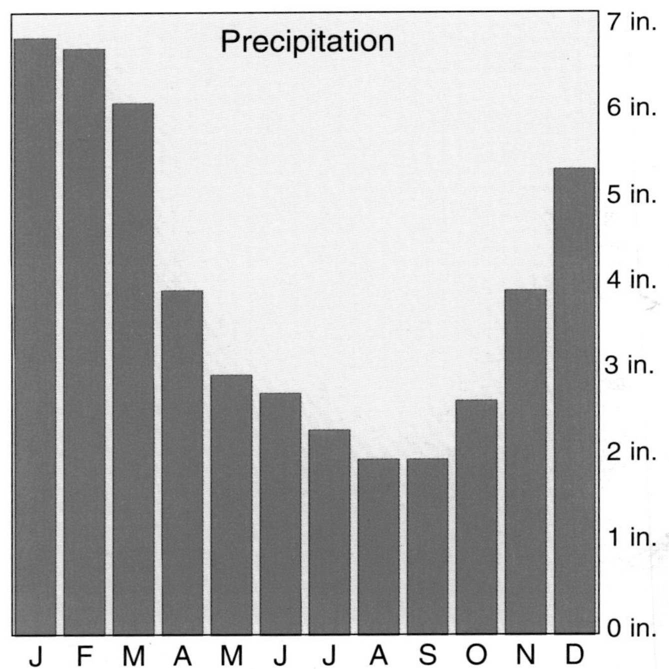

Brisbane's graphs show wet, warm summers.

Right *Berlin, Germany, has a rainy temperate climate with snowy winters.*

Below *Melbourne is in southeast Australia, which has a rainy, temperate climate with warm summers.*

Regions with temperate rainy climates and warm summers occur in much of Western Europe, including the British Isles. Here, the variable weather is caused by the influence of depressions and anticyclones, which occur throughout the year. Other temperate rainy regions with warm summers include southeastern Australia, western Oregon, and Washington state.

The third group of temperate rainy climates, with cool summers, includes western British Columbia in Canada, southwestern Norway, where temperatures are raised by the warm North Atlantic Drift, and southern Chile. Christchurch in New Zealand also has a climate of this kind.

RAINY WITH HOT SUMMERS

Temperate climates with hot summers and no dry season are ideal for the growth of forests. Winters are mild or even warm and snow is unusual.

Trees continue to grow throughout the year, so evergreens, such as cypresses, laurels, and evergreen oaks, are among the many trees found in these regions. Despite the cutting down of trees to create farmland, forests are still found in these regions.

In parts of the southeastern United States, there are mixed evergreen and pine forests, with pine forests along the sandy coastal plains.

The southeastern United States contains several swamp regions, one of which is the Everglades in Florida. This famous region has many thickly wooded islands covered by dwarf cypress and pine and mangrove forests near the coast. It supports three hundred kinds of birds, and six hundred kinds of mammals, including deer, otters, panthers, and raccoons. The king of the swamp is a reptile—the American alligator.

Southern Brazil, northern Argentina, and Uruguay also have a rainy climate with a hot summer.

Left *The hurricanes that bring rain to Louisiana also cause damage such as this.*

Top right *Alligators live in the Florida Everglades.*

Bottom right *In Suzhou Creek, Shanghai, China, the average temperature in the warmest month is 80°F.*

In these regions, forests probably once covered large areas, but they have been cut down or destroyed by fire. As a result, only a few patches of forest survive, and the region is mainly grassland.

East-central China has some evergreen forests in the south and mixed forests of evergreens and deciduous trees farther north. As in South America, large areas have been cleared for farming.

Because of the climate, warm-weather crops grow well in these regions. For example, the south-eastern United States produces cotton, sugarcane, and tobacco.

This region has also become a resort area, especially in the winter, when thousands of visitors arrive from the cold northeast.

RAINY WITH WARM SUMMERS

The British Isles and much of west-central Europe make up the largest region with a temperate rainy climate and warm summers. The highest temperatures occur in the summer, when high pressure brings hot air from the south and east(southwesterly winds are warmed by the Gulf Stream). In the winter, high pressure brings cold air from the north and east and sometimes causes long spells of freezing weather.

Most rain in Western Europe is from depressions that come from the Atlantic Ocean. The wettest and coolest areas are in the west, especially on highlands.

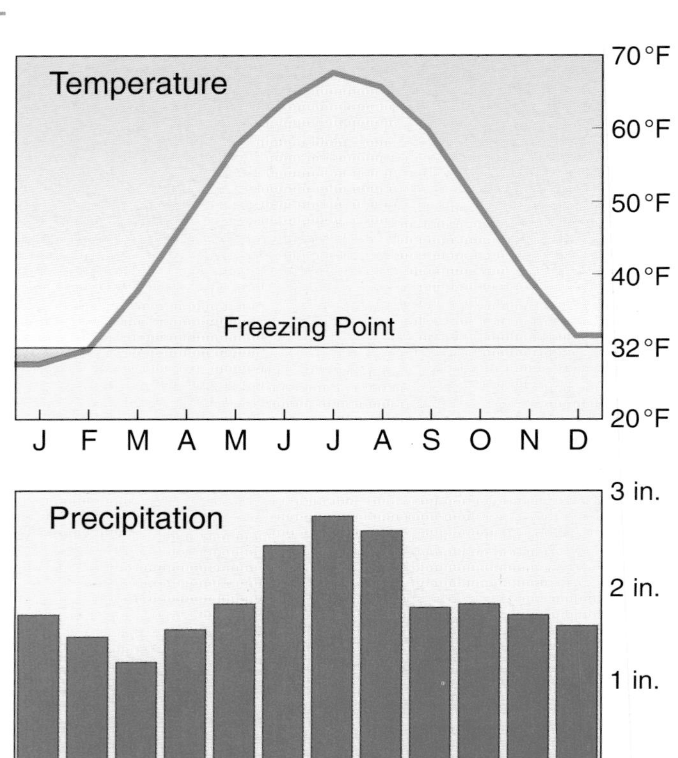

Climate graphs show Berlin's temperate rainy climate with warm summers.

Left *The marked differences between summer and winter in the British Isles meant that deciduous forests once covered large areas. Deciduous trees (ash, beech, birch, chestnut, and oak) shed their leaves in winter. Most of those forests have now been replaced by farms and grazing land.*

RAINY WITH COOL SUMMERS

Above *Christchurch, New Zealand, has a temperate rainy climate with cool summers.*

The third group of temperate rainy climates has cool summers. Regions with this type of climate lie in the high latitudes, but they are close to the sea, which moderates their climates. For example, southwestern Norway would have a much cooler climate were it not for the onshore winds that are warmed as they pass over the North Atlantic Drift, the northern extension of the Gulf Stream.

The climate of western British Columbia, Canada, is also affected by mild winds that blow onshore, bringing heavy rain. But the moderating effect is not so great as in southwestern Norway, and as a result western British Columbia has colder winters than southwestern Norway.

A third temperate region with cool summers is the southern part of Chile, where winds from the oceans bring storms, but they also moderate the temperature throughout the year.

The southeast coasts of South Island, New Zealand, have a similar temperate, though less wet, climate.

Cold Temperate Climates

CONTINENTAL MOIST CLIMATES

Some large areas in the middle latitudes have warm or hot summers but have winters that are cold and snowy. These places have a cold temperate climate. During the long winters, it is often as cold as in subarctic regions, and frosts are common in spring and fall. Because the frosts shorten the growing season, crops must be hardy to survive.

In summer, heatwaves often occur. They are sometimes severe. In July 1995, a heatwave in the north-central and the northeastern United States sent temperatures soaring above 100 °F (40 °C). Lasting nearly a week, it caused the deaths of more than 370 people in Chicago. The heatwave also claimed victims in New York City, Philadelphia, and Washington, D.C. Many of the people who died were elderly, and most of them lived in homes with no cooling systems.

Most regions with cold temperate climates lie far from the sea, in the hearts of continents. But some coastal areas, where cold ocean currents chill the onshore winds, also have cold temperate climates. For example, southeastern

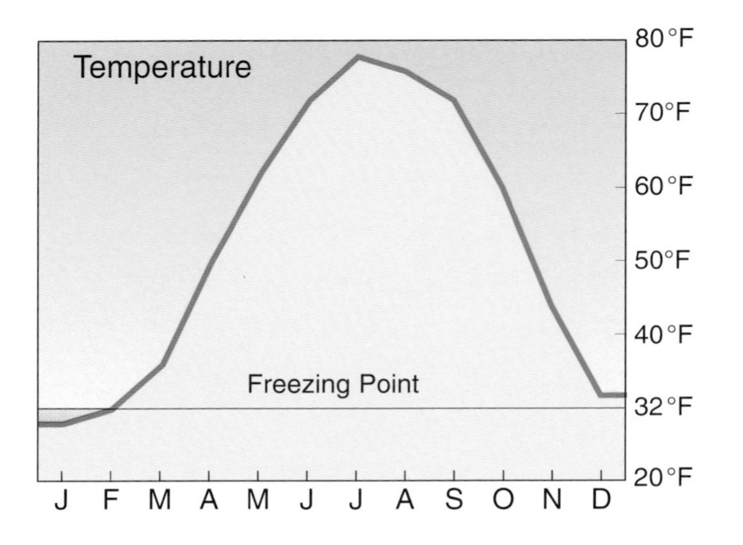

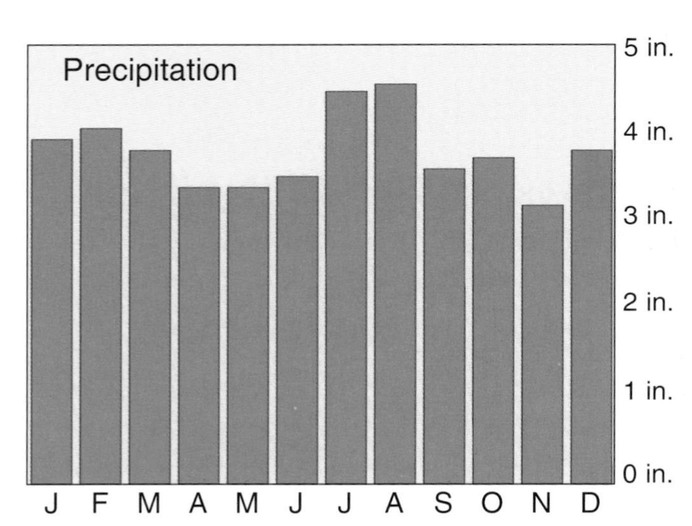

Climate graphs for New York City, where cold ocean currents reduce winter temperatures

Canada is cooled by the Labrador Current. The icy water in this current comes from the Arctic, often bringing jagged icebergs into the North Atlantic Ocean.

Other areas with cold temperate climates include the northeastern and midwestern United States, eastern Europe and west-central Russia, and northeastern China. The islands of Hokkaido and northern Honshu in Japan also have a cold temperate climate. These islands are chilled by the cold Oya Current, which comes from the Arctic.

No cold temperate regions occur in the Southern Hemisphere. That is because the higher middle latitudes south of the equator do not contain any large continental areas.

Above *The long, cold winters in Denver, Colorado, make it a popular resort for skiers.*

Right *Skaters on the Rideau Canal, Ottawa. Canada has a more severe climate than the northeastern United States, with a temperature range of 12° F (January) to 70° F (July).*

NORTH AMERICA

North America contains a large cold temperate region. It lies between the subarctic coniferous forest zone in Canada and the warm temperate region in the southeastern United States. It includes the Great Lakes region and the St. Lawrence River valley, New England, and New York State. In the Midwest this region merges into the dry prairie regions.

In the winter, the cold temperate region comes under the influence of a large high air-pressure system that tends to turn away depressions. This reduces the winter precipitation. Instead, the summer months are the wettest. Rain comes from depressions and also from thunderstorms, which form on hot

Above *The natural display of the autumn leaves in New England attracts many visitors.*

Left *The Great Lakes never completely freeze over. This is Lake Ontario.*

Right *Wheat is grown on the dry prairies of the Midwest.*

afternoons. Generally, the rainfall decreases from east to west. Boston, Massachusetts, has an average yearly precipitation of about 42 inches (1,040 mm), while Chicago, to the west, has 34 inches (840 mm).

The coldest parts of this region are in the north, but almost everywhere, average temperatures are below the freezing point for at least one month a year. The Great Lakes help moderate temperatures, and mild spells of weather often occur even in midwinter. However, there are also spells of bitterly cold weather when waves of polar air sweep south. These cold winds often bring fierce blizzards.

Unpleasant conditions also occur in summer. Heatwaves occur from time to time, when daytime temperatures may rise to 100 °F (38 °C) for several days.

Deciduous and coniferous trees once grew over much of this area, but most of the original forest has been cut down.

However, the mixed forests that grow in New England provide one of the world's finest spectacles. In autumn, the leaves of the deciduous trees change color, creating a blaze of yellow, gold, brown, scarlet, and pink, against a background of green from the coniferous trees.

WEST-CENTRAL RUSSIA

Another region with a cold temperate climate extends from eastern Europe through west-central Russia into Siberia. In the west, it includes southern Sweden, southern Finland, Estonia, Latvia, Lithuania, Belarus, and parts of Poland, Hungary, and Romania. In Russia, the region lies between subarctic, coniferous forests to the north and the dry steppes to the south.

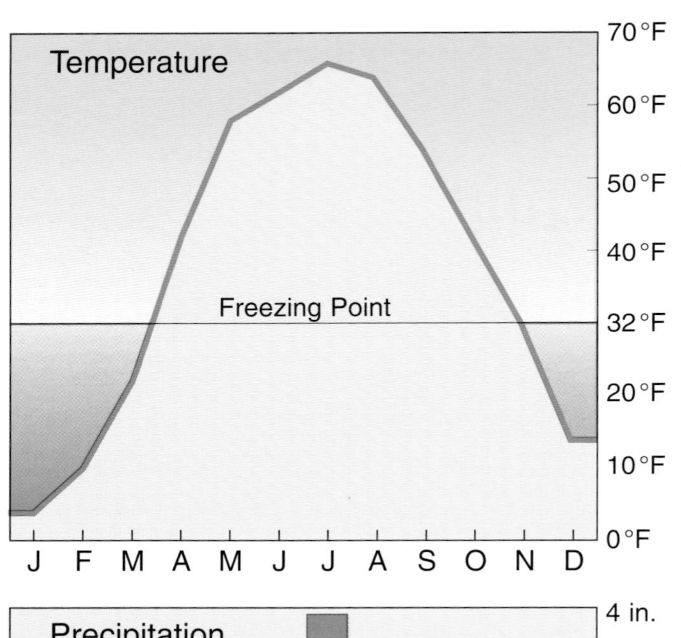

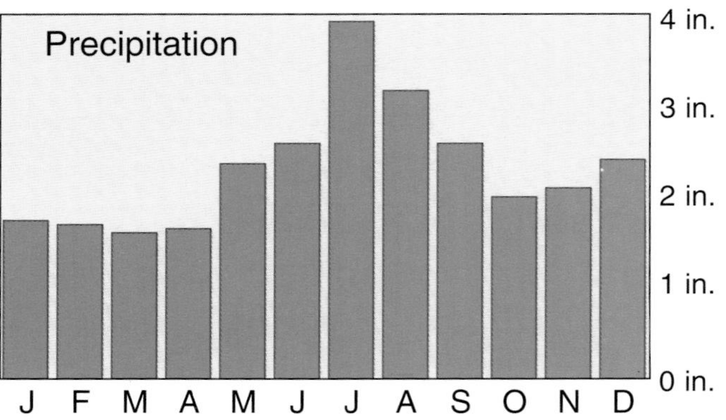

Climate graphs for Moscow show severe winters, warm summers, and moderate rainfall.

Winters are mildest around the Baltic Sea in the southwest. To the east, winters become increasingly severe. In west-central Russia winters are cold, much like those in southern Canada. For example, Moscow has five months when the average temperatures are at or below freezing. The cold winters have had an impact on Russia's history—they were an important factor in the defeats of Napoleon's French armies in the 1810s and Hitler's armies in the 1940s. In both cases, Russian troops were

Mixed forests of deciduous and coniferous trees grow in wetter parts of this region, with grasslands in drier areas. The forests once teemed with wildlife, but many animals have been hunted almost to extinction. The Bialowieza Forest on the border between Poland and Belarus gives a good idea of what these forests were like. It contains such trees as alder, ash, hornbeam, lime, oak, and pine and is the home of the rare wisent (European bison).

Left *Winters in St. Petersburg, Russia, are harsh and cold.*

Below *Rye, oats, and potatoes are crops that can grow during the short growing season of central European Russia. This grain silo for loading trucks is at Krasnodar.*

more able to cope with the harsh winters than their enemies.

Temperatures rise rapidly in spring, and summers are pleasantly warm, becoming warmer to the east and south. The precipitation is moderate, occurring throughout the year, although the largest amounts fall in the summer. In winter, the precipitation occurs mainly as snow. Even in Stockholm, in southern Sweden, snow covers the land for about 60 days every year. The summer rainfall comes from depressions and thunderstorms.

Changing Environments

CITY CLIMATES

Every place on earth has its own climate. Climatologists (people who study climate) divide the world into large climatic regions, including the temperate regions described in this book. However, even tiny areas, such as backyards, have their own climate. For example, a gently sloping yard facing south in the Northern Hemisphere has a slightly warmer climate than a sloping yard facing north. The study of the climate of small areas is called microclimatology—*micro* comes from the Greek word meaning "small."

Most people in temperate latitudes live in cities and towns. Many live and work in buildings that are centrally heated in the winter and air-conditioned in the

Left *Winds are channeled along this canyonlike street in New York.*

Right *Koshosdale power station in Kentucky releasing heat into the atmosphere*

Far right *An asphalt road melts in the heat. Asphalt absorbs more solar radiation than grasslands or forest. In a city, this absorbed heat from the roads can raise the air temperature.*

the summer. The insides of these buildings have an artificial climate.

Heating systems, together with cars and other vehicles, factories, and power stations, help to make cities warm places.

Also, the walls of buildings, stone pavements, and asphalt roads absorb heat from the sun. These surfaces then heat the air around them to a much greater extent than do the fields and farms around the cities. Hence, climatologists describe cities as "heat islands" set in the cooler "sea" of the surrounding countryside. In the winter, for example, cities often have fewer frosts than the surrounding countryside, and people in cities often have to pay less for their heating bills. The difference between the temperature in the city and the country is greatest on working days and smallest on weekends.

City buildings slow down winds blowing across the land. The moving air is sometimes funneled between buildings, producing strong gusts that can be very unpleasant.

CLIMATE AND POLLUTION

The activities of people in cities cause pollution, which affects climate. City factories and power stations burn fuels such as coal, gas, and oil. Smoke from factories and power stations contains a gas called sulfur dioxide. Other gases, called nitrogen oxides, are produced by gasoline engines.

In the air, sulfur dioxide and nitrogen oxides are dissolved by the tiny water droplets that make up clouds. This turns them into sulphuric and nitric acids. When the water droplets merge with others to form raindrops, they

Above *A coal-fired power station emits sulfur dioxide into the atmosphere.*

Left *Cars that run on gasoline emit fumes, including nitrogen oxide, into the atmosphere.*

Right *Acid rain has killed these forest trees in Bohemia in the Czech Republic.*

become acid rain. Because the clouds are blown by the wind, the acid rain often falls a long way away from the cities and factories. For example, industries in the northeastern United States may be responsible for acid rain in southeastern Canada.

All rain is slightly acidic because it has dissolved some carbon dioxide from the air, but acid rain is much stronger, and it is highly destructive when it reaches the ground. When it enters lakes and streams, it kills off fish and other living things. Acid rain also wears away stonework, such as the delicate stone carvings that decorate churches.

The most obvious effect of acid rain is the damage it has done to trees. Acid rain has destroyed large areas of forest in eastern North America, in northwestern and central Europe, and in parts of Asia.

Some countries have passed laws to try to reduce the amounts of sulfur dioxide and nitrogen oxides released into the air, which is the only way to reduce the amount of acid rain.

In some areas, lime is added to lakes and streams. Lime stops the harmful effects of the acid, but it is only a temporary measure, and it may have harmful side effects.

VANISHING FORESTS

In Germany, people say that trees attacked by acid rain are suffering *Waldsterben* ("forest death"). But acid rain has become an important factor in killing off trees only in the last 50 years. Most of the original forests of the temperate regions were destroyed long ago.

Above *Bialowieza National Park, Poland*

Above right *A cedar of Lebanon —only a few of these trees have survived, because their lumber was used to build ships.*

Above far right *A "clean air" demonstration against acid rain*

Left *Great Smoky Mountains National Park, North Carolina*

In Europe, in ancient times, farmers cleared the land to grow crops. They also used trees to build houses and ships. The destruction of forests was so widespread that very little original forest survived. For example, the New Forest in England was preserved only because it became a hunting area for Norman kings.

In what is now the northeastern United States, many Native American groups, such as the Chippewa, Huron, and Iroquois, lived in the mixed forests. They hunted, fished, and grew crops in forest clearings. When the soil in a clearing became infertile, the people moved on, and soon trees were again growing on the old clearing. But from the early seventeenth century on, European settlers began to develop the land. They cut down the forests, taking down trees for building and to create farmland.

The rich wildlife of the area dwindled because its forest habitats had disappeared.

The settlers exposed the land to the rain and wind. The rain flowed down bare hillsides, carving out deep gullies and removing the soil. This soil erosion made large areas as barren as the surface of the moon. The soil was washed into rivers, which became choked with mud. The swollen rivers burst their banks, causing floods.

It is only in the last 70 years that large-scale efforts have been made to control soil erosion and to conserve forests.

OUR CHANGING WORLD

Around 200 years ago, the world had about 1 billion people. Most of them lived in country areas and worked on the land. In the nineteenth century, more and more people moved into towns and cities to work in factories. This change, called the Industrial Revolution, was accompanied by a rapid increase in the world's population. By the mid-1990s, the world's population had reached 5.6 billion and the majority of people in temperate regions lived in cities and towns.

The factories and machines of the Industrial Revolution burned coal, oil, and natural gas. These are called fossil fuels because they are formed from the remains of once-living things.

When fossil fuels are burned they give off various gases. One of these gases is carbon dioxide, a "greenhouse" gas. It is called this because it traps heat that is

reflected back from the earth's surface, much as the glass in a greenhouse traps heat. The more carbon dioxide there is in the atmosphere, the more heat is trapped and the hotter it becomes.

Before 1700, carbon dioxide made up about 280 parts per million in the atmosphere. Nowadays it has risen to 355 parts and is still rising. As a result, world temperatures have risen by an average of from 0.5°F to 1°F (0.3–0.6°C). As a result, some of the world's ice has melted, and in turn the world's sea level has been rising by 0.072 inch (1.8mm) a year over the last 50 years. If this continues, many cities in temperate regions, such as Amsterdam and New York City, will eventually be flooded.

Human interference with nature is causing the world climate to change. These changes are also affecting the land and sea. Unless pollution is controlled, these changes are likely to increase in the future.

Left *The Thames barrier is designed to keep London from flooding at times of high tides and storm surges.*

Right *Polder land in the Netherlands has been reclaimed from the sea, but it will be lost if the sea level rises as a result of global warming.*

Glossary

Acid A sharp, sour substance that can dissolve many metals or minerals.

Atmosphere The layer of air around the earth.

Blizzard A storm in which strong winds blow powdery snow across the land.

Carbon dioxide A colorless, odorless gas found in the atmosphere. Green plants need carbon dioxide to live and grow. Animals breathe in oxygen and breathe out carbon dioxide. Carbon dioxide is released when coal, oil, and natural gas are burned.

Cloud A mass of tiny water droplets or ice crystals in the air. The droplets and ice crystals are so light that they do not fall to the ground.

Condensation The process by which invisible water vapor turns into a visible liquid form (as water) or a solid (as ice).

Equator A line of latitude running around the world exactly halfway between the North and South poles.

Equinoxes The two times of year when the sun is directly overhead at the equator, and when all places on earth have equal day and night —12 hours each.

Evergreen A shrub or tree that remains in leaf throughout the year.

Fahrenheit Scale used to measure temperature. The freezing point of water is 32 degrees Fahrenheit (32 °F), and the boiling point is 212 degrees Fahrenheit (212 °F).

Fossil Evidence of ancient life found in rocks.

Front The boundary between two masses of air with different characteristics, mainly between bodies of cold and warm air.

Frost An air temperature of 32 °F or below, at which water freezes. Frosts may kill plants.

Gully A deep channel worn by water, usually down a hillside.

Heath An area of uncultivated land covered by small shrubs.

Heatwave A period of unbroken

hot weather, when temperatures are well above average.

Hemisphere Half a sphere.

Hurricane A storm that forms over the oceans north and south of the equator. Other names for hurricanes are tropical cyclones, typhoons, and willy-willies.

Lines of Latitude Lines running around the earth parallel to the equator. They are measured in degrees between the equator (0° latitude) and the poles (90° north and south). Lines of longitude run around the world at right angles to the lines of latitude.

Ocean current A marked flow of water in the ocean. Surface currents are caused mainly by winds. Warm currents heat coastal areas. Cold currents lower temperatures.

Orbit The path followed by a heavenly body, such as the earth, around another body, such as the sun.

Pollution The fouling or harming of human, animal, or plant life.

Population The total number of people who live in a place.

Scrub A type of vegetation consisting of low shrubs or stunted trees.

Solstice The time in summer or winter when the sun is overhead at its farthest northern and southern points (the Tropic of Cancer and the Tropic of Capricorn).

Subtropical region A region between the tropics and the temperate regions.

Temperature The measurement of how hot or cold something is.

Tropics The zone between the Tropic of Cancer and the Tropic of Capricorn. In the center lies the equator.

Turbulent air Strong upward and downward currents in the atmosphere. This turbulence creates bumpy conditions for aircraft. Turbulence occurs at ground level when winds change direction and blow in a series of gusts.

Water vapor Invisible moisture in the atmosphere. Water vapor behaves like a gas.

Further Information

Books

Booth, Basil. *Temperate Forests.* Our World. Morristown, NJ: Silver Burdett Press, 1989.

Flint, David. *The Mediterranean and Its People.* People and Places. New York: Thomson Learning, 1994.

——*The World's Weather.* Young Geographer. New York: Thomson Learning, 1993.

Mason, John. *Weather and Climate.* Our World. Morristown, NJ: Silver Burdett Press, 1991.

Sayre, April P. *Temperate Deciduous Forest.* Exploring Earth's Biomes. New York: 21st Century Books, 1994.

Taylor, Barbara. *Weather and Climate*: Geography Facts and Experiments. Young Discoverers. New York: Kingfisher, 1993.

Tompkins, Terence. *Ravaged Temperate Forests.* Environmental Alert. Milwaukee, WI: Gareth Stevens, 1993.

Index